Farm Animals

Cows on the Farm

by Bizzy Harris

Bullfrog Books

Ideas for Parents and Teachers

Bullfrog Books let children practice reading informational text at the earliest reading levels. Repetition, familiar words, and photo labels support early readers.

Before Reading

- Discuss the cover photo. What does it tell them?
- Look at the picture glossary together. Read and discuss the words.

Read the Book

- "Walk" through the book and look at the photos. Let the child ask questions. Point out the photo labels.
- Read the book to the child, or have him or her read independently.

After Reading

- Prompt the child to think more. Ask: What did you know about cows before reading this book? What more would you like to learn about them?

Bullfrog Books are published by Jump!
5357 Penn Avenue South
Minneapolis, MN 55419
www.jumplibrary.com

Library of Congress Cataloging-in-Publication Data

Names: Harris, Bizzy, author.
Title: Cows on the farm / by Bizzy Harris.
Description: Minneapolis, MN: Jump!, Inc., [2021]
Series: Farm animals | Includes index.
Audience: Ages 5–8 | Audience: Grades K–1
Identifiers: LCCN 2020020995 (print)
LCCN 2020020996 (ebook)
ISBN 9781645277033 (hardcover)
ISBN 9781645277040 (ebook)
Subjects: LCSH: Cows—Juvenile literature.
Livestock—Juvenile literature.
Classification: LCC SF197.5 .H37 2021 (print)
LCC SF197.5 (ebook) | DDC 636.2—dc23
LC record available at https://lccn.loc.gov/2020020995
LC ebook record available at https://lccn.loc.gov/2020020996

Editor: Eliza Leahy
Designer: Molly Ballanger

Photo Credits: TODD17/Shutterstock, cover (left); Shutterstock, cover (right); Tim Gainey/Alamy, 1; Eric Isselee/Shutterstock, 3, 14, 22; Supawinee Prachum/Dreamstime, 4; Photoagriculture/Shutterstock, 5, 18, 23tm; WitthayaP/Shutterstock, 6–7, 23tr, 23bl; JackF/iStock, 8–9; VanderWolf Images/Shutterstock, 10–11, 23bm; Clara Bastian/Shutterstock, 12–13, 16–17; Dajra/Shutterstock, 13, 23tl; Ben Schonewille/Shutterstock, 15; Nancy Gill/Shutterstock, 19, 23br; James Stevenson/Dreamstime, 20–21; New Africa/Shutterstock, 24.

Printed in the United States of America at Corporate Graphics in North Mankato, Minnesota.

Table of Contents

Milking and Chewing

Moo!

The cows are awake.

It is time to milk them!
The farmer herds them.

They go to the milking parlor.

Machines milk them.

milking
machine

hay

The farmer brings hay.

The cows eat.

Yum!

Cows eat grass, too.

Where?

In the pasture!

pasture

Cows swallow food.

It comes back up as cud.

They chew the cud.

Calves are young cows.

They stay with their moms.

tail

A bull is a male.

His tail swishes.

It flicks flies away.

The cows are herded.

Where?

To the barn!

They go into their stalls.

It was a long day.
The cows sleep.
Good night!

Parts of a Cow

Take a look at the parts of a cow!

Picture Glossary

cud
Food that cows bring up from the first part of their stomachs to chew again.

herds
Moves animals together in a group.

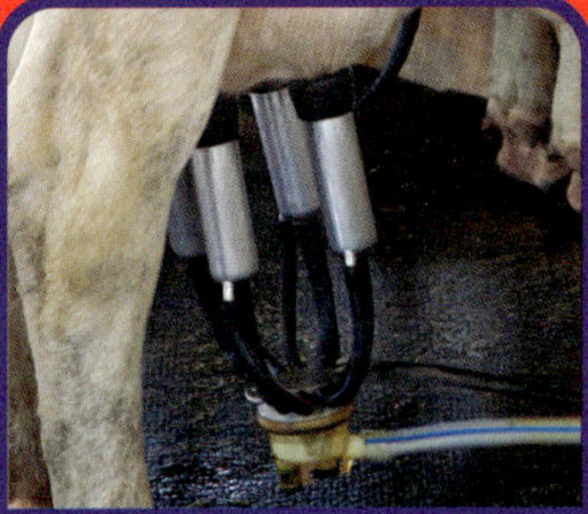

milk
To take milk from a cow or other animal.

parlor
A separate space where cows are led for milking.

pasture
Grazing land for animals.

stalls
Enclosed spaces for animals in a barn or stable.

Index

To Learn More

Finding more information is as easy as 1, 2, 3.

1. Go to www.factsurfer.com
2. Enter "cowsonthefarm" into the search box.
3. Choose your book to see a list of websites.